This Walker book belongs to:

For my dearest Rikka

First published 1992 by Walker Books Ltd
87 Vauxhall Walk, London SE11 5HJ

This edition published 2016

2 4 6 8 10 9 7 5 3 1

This book has been typeset in Garamond

Printed in China

British Library Cataloguing in Publication Data:
a catalogue record for this book is available from the British Library

ISBN 978-1-4063-4979-5

www.walker.co.uk
jezalborough.com

WHERE'S MY TEDDY?

JEZ ALBOROUGH

WALKER BOOKS
AND SUBSIDIARIES
LONDON • BOSTON • SYDNEY • AUCKLAND

Eddy's off to find his teddy.
Eddy's teddy's name is Freddy.

He lost him in the wood somewhere.
It’s dark and horrible in there.

"Help!" said Eddy. "I'm scared already!
I want my bed! I want my teddy!"

He tip-toed
on and on
until …

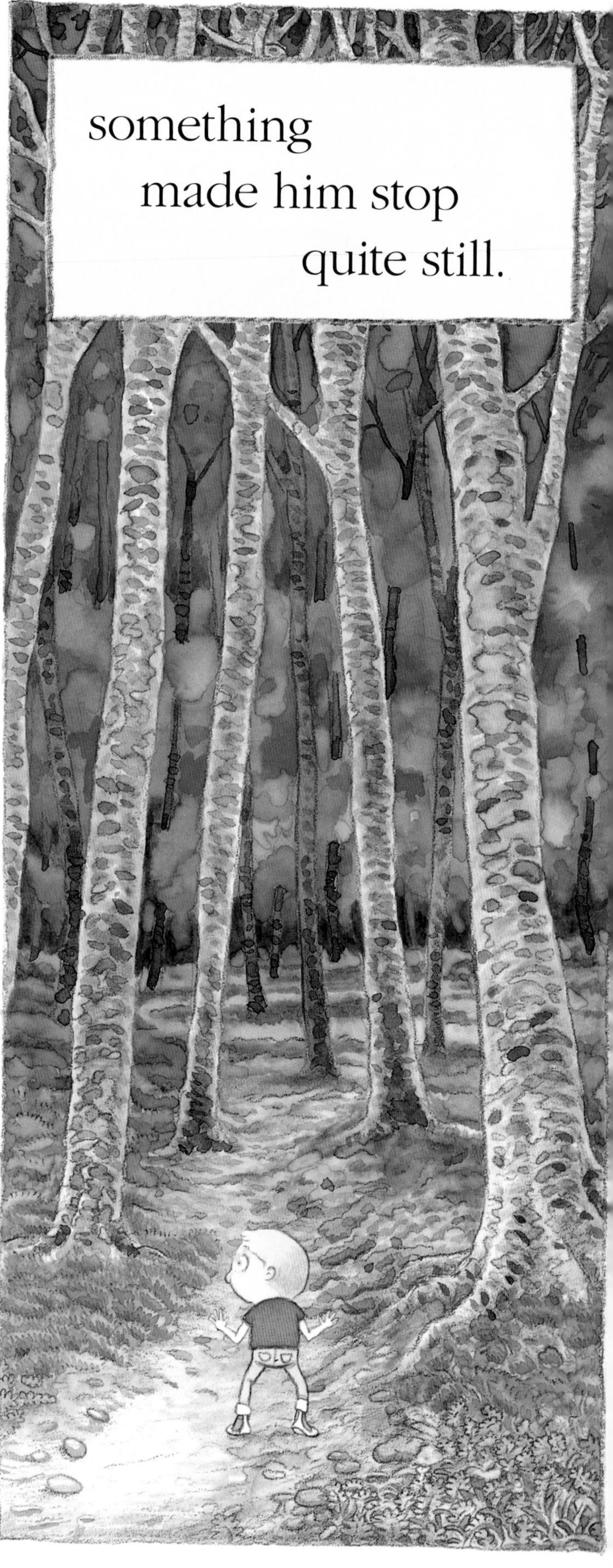
something
made him stop
quite still.

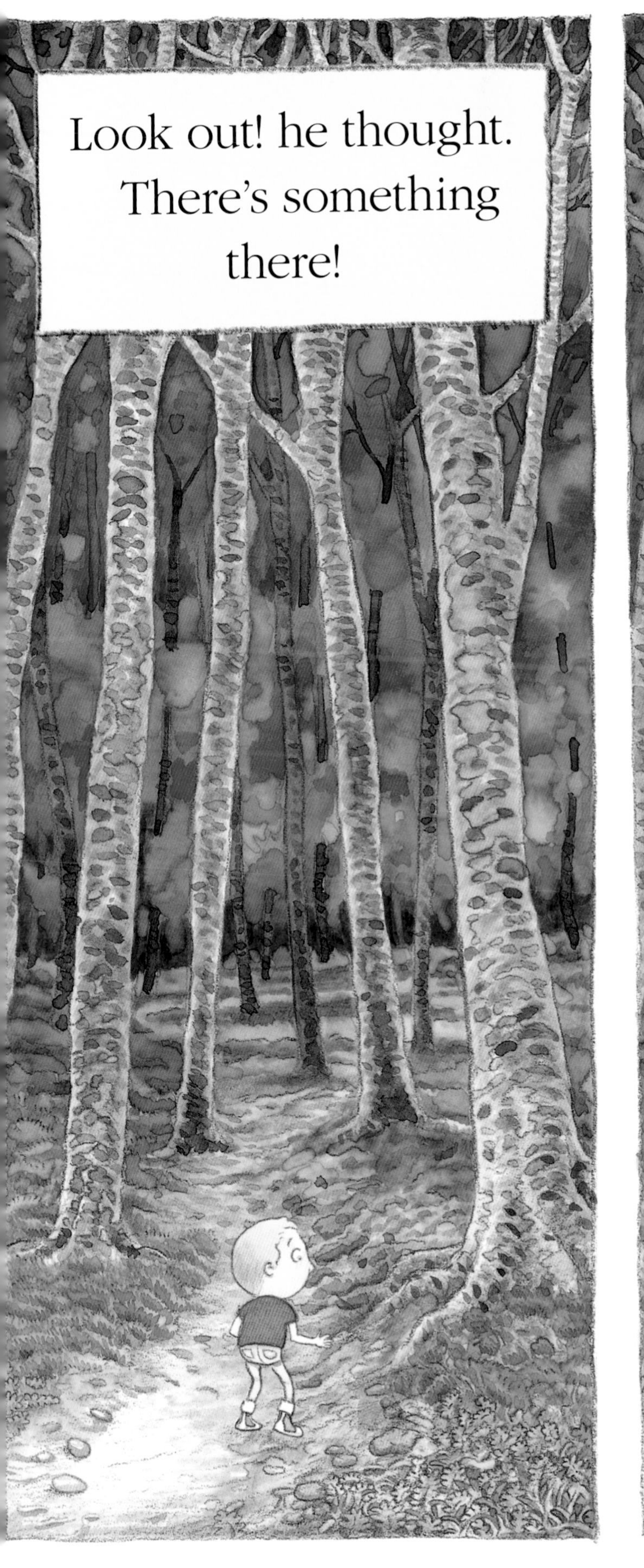
Look out! he thought. There's something there!

WHAT'S THAT?

A GIANT TEDDY BEAR!
"Is it Freddy?" said Eddy.
"What a surprise!
How *did* you get to be this size?"

"You're too big to huddle and cuddle," he said,

“and I’ll never fit both of us into my bed.”

Then out of the darkness,
clearer and clearer,
the sound of a sobbing
came nearer and nearer.

Soon the whole wood
could hear the voice bawl,
“How did you get to be
tiddly and small?
You’re too small to
huddle and cuddle,” it said,
“and you’ll only get lost
in my giant-sized bed!”

It was a gigantic bear
and a tiddly teddy
stomping towards …

the giant teddy and Eddy.

"MY TED!"
gasped the bear.
"A BEAR!"
screamed Eddy.

“A BOY!”
yelled the bear.
“MY TEDDY!”
cried Eddy.

Then they ran and they ran
through the dark wood
back to their homes
as quick as they could …

all the way back
to their snuggly beds,
where they huddled
and cuddled their
own little teds.

BOOKS BY JEZ ALBOROUGH

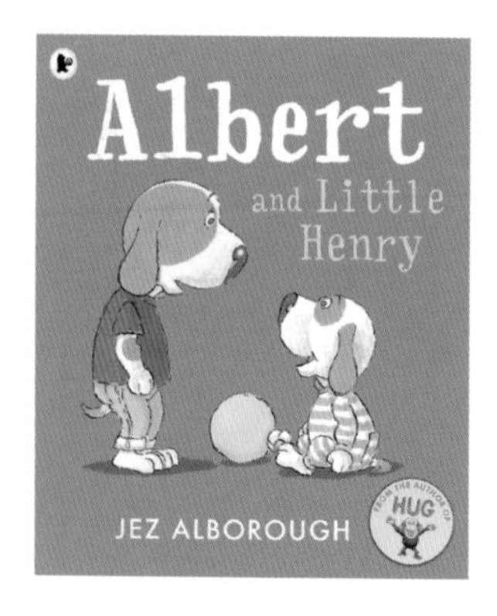

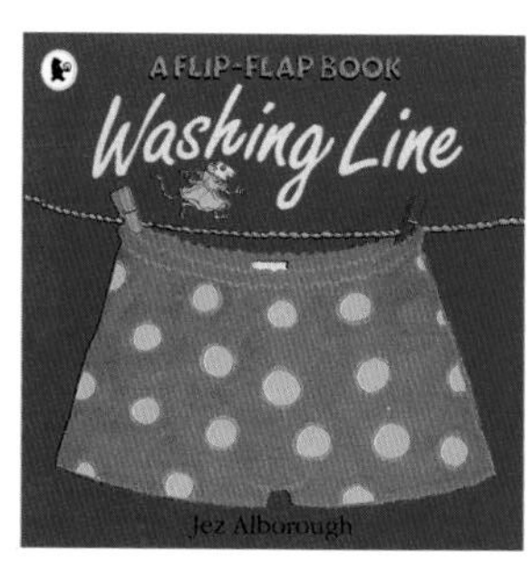

JezAlborough.com

Available from all good booksellers

www.walker.co.uk